YOUR KNOWLEDGE HAS VALUE

- We will publish your bachelor's and master's thesis, essays and papers

- Your own eBook and book - sold worldwide in all relevant shops

- Earn money with each sale

Upload your text at www.GRIN.com
and publish for free

Biosynthesis and Development of Novel Method for Commercial Production of Biosurfactant Utilizing Waste Potato Peels

Unnati Patel

Bibliographic information published by the German National Library:

The German National Library lists this publication in the National Bibliography; detailed bibliographic data are available on the Internet at http://dnb.dnb.de.

ISBN: 9783346970510
This book is also available as an ebook.

Biosynthesis and Development of Novel Method for Commercial Production of Biosurfactant utilizing waste potato peels

Running title- BIOSYNTHESIS OF BIOSURFACTANT FROM WASTE POTATO PEELS

Unnati Patel

Sanjivani College of Pharmaceutical Education and Research, Kopargaon, Maharashtra-423603, India

Abstract

Biosurfactant or microbial surfactants are surface dynamic atoms that are delivered by an assortment of microorganisms including microscopic organisms, yeast and filamentous parasite. In contrast with engineered synthetic surfactants, bio-surfactants are of interest because of their high level of activity, particular movement at outrageous temperatures, pH and saltiness, capacity to be delivered from inexhaustible feedstock and high level of biodegradability. In any case, bio surfactants have not yet been marketed widely because of low creation yields and high feedstock and cleansing expenses. With a specific end goal to tackle this issue many investigations have been done utilizing minimal effort feedstock or farming side-effects as substrates for bio surfactant creation. The foremost extent of the test work was accordingly; to create compelling strategy, and sanitize the biosurfactant generation by waste potato peels and pulp utilizing mutated *B.Subtilis* DDU20161. Major upstream parameters like creation of medium, fomentation, and pH were kept up in the investigation. The ideal biosurfactant yield by acetone extraction technique was 253.79 mgL-1 with Biomass of 3.56 gL-1 at 40 hrs. The produced Biosurfactant was characterised by studying surface tension, emulsification index, TLC, CMC and FTIR study. The highest emulsification index was 75.05% in Benzene, Critical Micelle Concentration(CMC) was found 0.028µM, and FT-IR study revealed presence of peptide bond at 3188cm^{-1} (=N-H). The outcomes acquired are the sole properties of biosurfactant according to announced literatures. The results of stability study indicated that the produced biosurfactant is steady at 30°-100⁰C without any significant change in surface tension properties of biosurfactant. The shaped biosurfactant is also found to be steady in alkaline pH. The present work is concentrate on basic, temperate and vast scale creation of biosurfactant utilizing acetone precipitation system.

Key Words- Biosurfactant, *B.Subtilis* DDU20161, M9 salt solution, Potato pulp and Potato peel as substrate

Introduction

Biosurfactants are amphipilic compounds that confer the ability to accumulate between fluid phases, causing reduced surface and interfacial tension of the samples[1]. They are a structurally diverse group of surface-active molecules synthesized by microorganisms[2]. Biosurfactant have attracted much attention in recent years; they are surface-active compounds synthesize by microorganisms that exhibit diverse chemical structures, including lipopetieds, polysaccharide-protein complexes, phospholipids, fatty acids and neutral lipids. Due to their amphipathic nature, biosurfactant display a variety of surface activities, which allows their application in several fields related with emulsification, foaming, detergency, wetting, dispersion and solubilization of hydrophobic compounds. Biosurfactants are produced by a wide variety of bacteria, yeast, and filamentous fungi[3]. *Bacillus subtilis* has been used in the biological control of plant diseases. It secretes a broad spectrum of bioactive peptides, with great potential for biotechnological and biopharmaceutical application[4].Many biosurfactant have been reported to possess a similar or better performance when compared with synthetic surfactant, which in addition to their lower toxicity, higher biodegradability and effectiveness at extreme temperatures, salinities and PH values, make them a green alternative to their chemical counterparts in different application, including agriculture, food, cosmetics or petroleum industries, as well as in bioremediation[5]. However, biosurfactants have not yet been commercialized extensively due to low production yields and high feedstock and purification costs. Therefore hopefully tomorrows microbial surfactants appear to depend particularly on the use of plentiful and cheap substrates for optimization of the operational cultivation condition, which can markedly increment the yield[6].

Most Bacilli and some other species of bacteria have a wide range of hydrolytic enzyme systems and are often capable of utilizing the organic matter consisting of complex mixtures. These organisms are easy to grow and require no expensive growth factors. The Bacillus will also be tried for the extraction of extra cellular enzymes & biosurfactant from potato peels. The wastes coming out from potato plants in the form of liquid and solid arise from peeling, trimming, slicing, cleaning, and rinsing operations which creates pollution problem. Looking to the above it was thought to utilize the byproduct such as potato peels of the potato industries for the production medium· of biosynthesis of biosurfactant and also to solve the problem of pollution. There are some review are done for same case[3-7].

Increasing biosurfactant yields and decreasing production costs are essential factors affecting the efficiency of biosurfactant production process found that the use of costly substrate, gave low yields and accumulation of undesirable product mixtures rather than refined biosurfactant compounds, such constraints explain why there is restricted production of biosurfactant in industry [8]. In order to solve this problem many studies have been carried out using low-cost feedstock or agriculture by products as substrates for biosurfactant production. The present study was aimed on the use of waste potato peels and pulps as a substrate, effective solvent system for the extraction of biosurfactant to maximize biosurfactant yield on large scale.

Materials and Method

Microorganism and culture preservation

B.Subtilis bacteria was isolated from oil contaminated soil nearby soap industry, Kheda Highway, Kheda, Gujarat and identified by the rRNA 16S sequence. The isolated, identified and mutated organism was denoted name as *B.Subtilis* DDU12016. Bacteria was cultured in fresh N-broth liquid media and the culture was kept on slants, refrigerated at 4°C and replicated every 30days.

Experimental Conditions for the Production of Biosurfactants

For synthesis of biosurfactant, the culture medium was prepared according to Moran *et al.* (2000), with slight modifications. *B.subtilis* DDU20161 was grown on minimal media, nutrient salt solution and potato as substrate. Composition of minimal media and culture conditions as follows:

Bacterial culture: Fresh culture of B.*subtilis* DDU20161 was prepared by using a nutrient broth as culture media. Composition of Nutrient broth as peptone 1gm, beef extract 1 gm, sodium chloride 0.5gm in distilled water. A loopful culture of B.*subtilis* DDU20161 from Nutrient agar slant transpired into a Nutrient Broth solution it incubated for 24hrs at 37^{0}C.

Minimal Media: Minimal Media basically composed of M9 salt solution, sugar, ammonium chloride, Ions, sodium, potassium, calcium. Composition of M9 salt solution: sodium phosphate (64gm), Potassium Phosphate (16gm), Ammonium Chloride (5gm) dissolve in water and make up to 1000ml then sterilize by using autoclave or vacuum filtration. To prepare 1L of minimal media add above prepared 200ml of M9 salt solution to 800ml distilled water, then autoclave it and add 2 ml of MgSO4 and 20 ml of 20% glucose.

Nutrient solution: Nutrient solution was prepared by using glucose (carbon source) and Ammonium Chloride (Nitrogen source) within ratio of 1:1.

Propagation of the Inoculum

The microorganism was replicated in a Petri plate containing minimal medium along with Agar (15gm/1L), and incubated at 30°C for 48 hours. In a 500ml of Erlenmeyer flasks, added 50 ml of minimal media, 10 ml of nutrient solution (Glucose: NH_4Cl; 1:1) mix it well. To this Potato: Potato peel substrate (4:4gm) was added and mix it homogeneously. Maintain pH of media 7 by using 1M NaOH solution. At the end transferred 2-3 ml of bacterial culture under aseptic condition to flask. Incubate for 48 hrs at 37^0C in rotary shaker at 180 rpm.

Production of Biosurfactants in a Continuous Stirred Tank Bioreactor

The production of biosurfactant was carried out in a 5-L bioreactor (BIOSTAT Cplus; Sartorius) operating with 3 L of available volume, using the culture medium. A concentration of 15% v/v of inoculum (0.95 g/L) was used and the fermentation was conducted at 37°C, 180 rpm and with 1 L/min aeration. The samples were collected at pre-defined intervals to determine biomass and later centrifuged at 8000 rpm for 20 minutes at 4 °C to remove the cells. Clear supernatant was observed which served as source of crude biosurfactant. Biosurfactant was recovered from the cell free culture supernatant by cold acetone precipitation, ammonium sulphate precipitation and acid precipitation methods[8]. Equal volume of chilled acetone added to supernatant; allow standing for 10 hrs at 4^0C. Precipitate was collected by centrifugation and evaporated to dryness to remove residual acetone after which it was re-crystallizing with chloroform and methanol[9]. The precipitate of biosurfactant was then analyzed for biosurfactant concentration and characterization. Moreover, the biosurfactant was also extracted by using ammonium precipitation and acid precipitation methods for comparision of biosurfactant extractive yields[10].

Characterization of isolated biosurfactant

Biomass Yield

Biosurfactant production has been associated with cellular growth of bacteria. Biomass yield of bacteria was determined through turbidometry, using a spectrophotometer (Shimadzu, 1607) at 660 nm, where biomass (g/L) was determined using a calibration curve of biomass vs optical density[11].

Analysis of amino acid

The Biosurfactant isolated from *B.Subtilis* generally categories as lipopeptide type of Biosurfactant. Physical test such as appearance of solution, color of solution, litmus test and Biurret test were performed for presence of protein.

Biosurfactant Concentration

The concentration of biosurfactant was determined by High Performance Liquid Chromatography, coupled with a UV detector (Shimadzu, model 228) at 205 nm and equipped with a Phenomenex Luna 5u C18 column (250 x 4.6 mm, 5 μm, Sl. No. 534094-1). The composition of the mobile phase was 20% of trifluoroacetic acid (3.8 mM) and 80% of acetonitrile. The flow rate was 1 mL/min at 30 °C and the volume of sample injections was 20 μL. The samples were analysed using a calibration curve prepared with standard surfactin (99% pure) from Sigma-Aldrich[12]. The concentration of biosurfactant was calculated from the chromatographic peaks, according to Wei and Chu (1998).

FT-IR analysis

FT-IR spectroscopy of isolated biosurfactant was carried out for the confirmation of isolated surfactant and commercially available surfactant. The IR spectra were recorded on FTIR-8400shimadzu spectrometer, in 4000 to 400 cm^{-1} spectral region. All three isolated powder sample from respective three media were analyzed by FT-IR analysis.

Surface tension measurement

Surfactant helps in reducing surface tension and the interfacial tension. The surface tension measurement of cell free supernant was determined using stalagnometer by drop weight method at 25^0C. Surface tension cell free culture isolated from nutrient media was calculated using following equation.

$$Y2 = \frac{Y1\,M2}{M1}$$

Y1=surface tension of water

Y2= surface tension of sample

M1=mean wt of one drop of water

M2= mean wt of one drop of sample

Emulsification Index (E24)

E24 of culture samples was determined by adding 2 ml of hydrocarbon (Benzene, Toluene, and Hexane) to the same amount of cell free supernatant, mixing with a vortex for 2 min, and leaving to stand for 24hrs. The E24 index is given as percentage of height of emulsified layer (mm) divided by total height of the liquid column (mm)[9].

$$E24 = \frac{height\ of\ emulsion\ layer}{total\ height\ of\ the\ liquid\ column} \times 100$$

Foaming activity

The foam was produced by hand shaking a cell free broth of biosurfactant solution for a several minutes. Stability of foam was monitored by observing it during an hour.

TLC study

Preliminary characterization of biosurfactant was done by Thin Layer Chromatography (TLC). Isolated biosurfactant were separated on silica gel plate (TLC plate) using chloroform: methanol: water (65:15:1)[8]. The component were detected by spraying the plate with distilled water and observed under UV-cabinet (Figure 4).

Critical Micelle concentration of isolated surfactant was determined spectrophotometrically by using curcumin as probe. Curcumin is previously used as a probe for determination of protein in actual samples. *Satyajit Mondale* et al.[13] first time employed curcumin for determination of critical micellar concentration using UV absorption spectroscopy. A stock solution of curcumin (1mg/ml in methanol) was prepared by adding a known weight of compound in methanol solution. A small quantity of curcumin stock solution 8.4μl was added to 2.5 ml of water to achieve a final concentration of 9.1μM for curcumin. The concentrated surfactant solution was stepwise added into aqueous curcumin solution and absorbance was recorded at Λmax 368nm after each concentration on UV-1650 shimadzu.

Stability study

For determination of the thermal stability of the biosurfactant, cell-free broth was placed at a constant temperature range of 20–100 °C for 15 min, then cooled to room temperature and activity of the biosurfactant was investigated by measuring the surface tension. To study the effect of pH on activity, the pH of the cell free broth was adjusted to different values using 1N NaOH or 1N HCl. The effect of addition of different concentration of NaCl on the surface tension of the biosurfactant was investigated. The biosurfactant was redissolved after purification with distilled water containing the specific concentration of NaCl (0–9%, w/v)[14-15].

Result and Discussion

Growth of bacteria and production of biosurfactant is associated with medium composition. Total carbohydrate assays and glucose monitoring indicated that *B. subtilis* was able to degrade potato substrates to produce surfactant[15]. Table 1 shows profiles of Biomass and Biosurfactant production. From tabulated data, it was observed that the culture reached the stationary phase after 40 hours of cultivation, with a maximum concentration of 3.56 g/L in 40 hours. Observing the production of biosurfactant, it can be seen that the concentration of biosurfactant increased during the microorganism's exponential phase. The production of biosurfactant by *Bacillus subtilis* reached a maximum value of 253.79 mg/L in 40 h of cultivation, while a decline in product formation after 50 h of fermentation was observed. It is important to mention that, during this process, an important amount of foam is produced. Some of researcher[11] have reported a maximum concentration of biosurfactant of 232 mg/L in 48 h, which remained constant until 72 h, using a mineral medium supplemented with glucose and fructose. In comparision with above results from the reported literature, the production of biosurfactant by *Bacillus subtilis* using potato peel and potato pulp was more efficient, since higher concentrations of biosurfactant were obtained. Kim *et al.* (1997) studied the production of biosurfactant by *Bacillus subtilis* C9 in a mineral and glucose medium and reported that the formation of product was associated with cellular growth.

Amino acid analysis of biosurfactant was carried to confirm the presences of protein. Table 2 showed test positive indicated that isolated biosurfactant posses protein structure and it might be a lipopeptide type of surfactant (Table.

FT-IR analysis carried to confirm that isolated surfactant and commercially available surfactant having same functional group or not. The IR spectra given below showed the similar functional group as available surfactant (Table 3).

Figure 2 shows FT-IR spectra of marketed surfactin (Sigma-Aldrich) compared with the biosurfactant produced by using *Bacillus subtilis*. This study revealed that the biosurfactant produced presents the main functional groups of a surfactin molecule, indicating the presence of aliphatic hydrocarbon, as well as a peptide component.

The critical absorption bands were assigned by comparison with spectra obtained from the literature (Lin *et al.*, 1994). In Figure, seven main bands can be observed. The absorption band with a maximum of 3307 cm^{-1} corresponding to the N-H stretch showed presence of peptide residues. In second case band with maxima of 2958 cm^{-1} and 2927 cm^{-1}, corresponding to the C–H stretch, which might be because of the lipopeptide portion of biosurfactant. At 1726 cm^{-1}, a small intensity band is observed that can be due to the absorption of C=O groups. At 1651 cm^{-1}, a

CO–N stretch points to the amide group. The bands at 1466 cm^{-1} and 1388 cm^{-1} indicate aliphatic chains. These results suggest that the biosurfactant produced by *Bacillus subtilis* is a cyclic lipopeptide, by and large surfactin.

Biosurfactants obtained from *Bacillus subtilis* are classified into three families of cyclic compounds: iturin, surfactin, and fengycin. It has been reported previously that derived biosurfactant has very good surface active properties (Raaijmakers *et al.*, 2010). So, our claim for confirmation that the produced biosurfactant was surfactin by the use of different techniques likes HPLC and FT-IR. HPLC study showed that chromatogram of the biosurfactant sample produced by *Bacillus subtilis* DDU20161 showed similar retention peaks (1, 2, 3, 4, 5 and 6) to those observed with marketed surfactin (Figure 3).

Surface tension analysis of isolated Biosurfactant on an average had 29.96mN/M^2 while surfactin showed 27.65mN/M^2.

Emulsification index (E24%) of produced biosurfactant was studied in various organic solvents. Results of emulsification study showed E24% in Benzene (75.05%), Toluene (65.12%) and Hexane (42.36%) subsequently. However, the emulsification index has been affected by number of factors like temperature, pH, salinity etc. The detail stability study results were incorporated considering effect of temperature and pH on emulsification index (E24%).

Preliminary characterization of biosurfactant using TLC revealed a white dry spot with retention index of 0.475 (Fig. 4); the standard surfactin presented a retention index of 0.55[10].

CMC value was determined by spectrophotometrically by using curcumin as probe. In this method addition of surfactant solution into standard curcumin solution gave absorbance in increasing order upto certain extent (Table 4 & Fig.6). While from CMC point and above, it showed the constant absorbance (This point was considered as CMC value of biosurfactant) at 368 nm. The CMC value was found to be 0.028µM.

It is an important to study the working ability of biosurfactant under different temperature conditions. The stability of biosurfactant was tested over a wide range of temperature. The biosurfactant produced by *B.subtilis* DDU20161 was shown to be thermo-stable. Heating of the biosurfactant to 100°C caused no significant effect on the biosurfactant performance. The emulsification index was also found to be stable under the study temperatures used (E24=75%) in comparison with synthetic surfactants such as SDS which exhibits a significant loss of emulsification activity beginning at 70 °C[16]. Therefore, it can be concluded that this biosurfactant maintains its surface properties unaffected in the range of temperatures between 30 and 100 °C. This activity was discovered indicating the usefulness of the biosurfactant in food, pharmaceutical and cosmetics industries where heating to achieve sterility is of prime importance[17-18].

The surface tension of the isolated biosurfactant remained stable to pH changes between pH 8 and 12 and showed higher stability at pH 9 than acidic conditions. At pH 12, the value in emulsification activity (E24) showed almost 63%, whereas below pH 7 it was decreased up to 40%. In addition, for pH values lower than 5, the solution of biosurfactant become turbid, due to partial precipitation of the biosurfactant. Study results indicate that increase pH has a positive effect on emulsification activity and emulsion stability. This could be caused by a better stability of fatty acid surfactant micelles in the presence of NaOH and the precipitation of secondary metabolites at higher pH values. The effect of pH on surface activity has been reported for biosurfactants for different microorganisms[19].

Conclusion

Biosurfactant generation endeavor was done utilizing B. subtilis and potato substrate as a decision of medium. UV transformed species alongside potato pulp and potato peel (3:1) demonstrated great biosurfactant development in bacterial cell because of presence of some particular salts and electrolytes in potato peel waste. Biosurfactant creation was observed to be ideal at 40 hrs with generation 253.79mgL-1of biosurfactant with 3.564 gL-1 biomass arrangement. FT-IR study revealed similarity between isolated biosurfactant with commercially available surfactin. Emulsification index was observed to be optimum in Benzene (75.05%). Results of stability study under different temperature conditions showed isolated biosurfactant was unaffected by temperature under study. Moreover, the produced biosurfactant is also found to be stable in pH 8 to 12 and more active in alkaline condition (pH 9). So, above all results indicates commercial production viability with wide industrial applicability.

Acknowledgment

The authors are highly thankful to Sanjivani College of Arts, Commerce and Science, Kopargaon, Maharashtra for providing generous support in media and other necessary facilities to carry out this work.

References

1. Muthusamy K, Gopalakrishnan S, Ravi TK & Sivachidambaram P, Biosurfactants: Properties, commercial production and application, *Current Science, 94* (2008) 736-747.

2. Lu J, Zhao X & Yaseen M, Biomimetic amphiphiles: Biosurfactants, *Curr Opin Colloid Interface Sci,* 12 (2007) 60–67.

3. Nurullah A & Fikret U, Production of extracellular alkaline protease from Bacillus subtilis RSKK96 with solid state Fermentation Eurasia, *J Biosci*, 5 (2011) 64-72.

4. Ashlee ME & Richard L, Ecology and genomics of *Bacillus subtilis, Trends Microbiol*, 16 (2008) 269-275.

5. Muhammad I & Muhammad N, Production of Thermostable α-amylase from Bacillus Sp. In Solid State Fermentation, *JASR*, 7 (2011) 607-617.

6. Mahmood AU, Greenman J & Scragg AH, Orange and potato peel extracts: Analysis and use as *Bacillus* substrates for the production of extracellular enzymes in continuous culture. *Enzyme Microb. Technol,*22 (1998)130-137.

7. Tamires CDS & Devson PPG, Optimisation of solid state fermentation of potato peel for the production of cellulolytic enzymes, *Food Chem*, 133 (2012) 1299–1304.

8. Marcia N & Glaucia M P, Production and properties of a surfactant obtained from Bacillus subtilis grown on cassava wastewater, *Bioresource Technol*, 97 (2006) 336–341.

9. Abouseoud M, Biosurfactant Production from olive oil by Pseudomonas Fluorescens Communicating Current Research and Educational Topics and Trends in Applied Microbiology A, Méndez-Vilas (Ed.) 2007, 340-47.

10. Lee SC, Kim SH, Park I H, Lee YS, Chung SY & Choi YL, Characterization of new biosurfactant produced by *Klebsiella* sp. Y6-1 isolated from waste soybean oil. *Bioresource Technol*, 99 (2008) 2288-2292.

11. Giro, M E, Martins A, Rocha J L, Melo KL & Gonçalves LRB, Clarified cashew apple juice as alternative raw material for biosurfactant production by *Bacillus subtilis* in a batch bioreactor, *Biotechnol J*, 4 (2009) 738-747.

12. Yeh CC, Hou MF & Tsai SM, Superoxide anion radical, lipid peroxides and antioxidant status in the blood of patients with breast cancer, *Clin Chim Acta*,361 (2005)104–11.

13. Satyajit M & Soumen G, Role of curcumin on the determination of the critical micellar concentration by absorbance, fluorescence and fluorescence anisotropy techniques, *J Photochem Photobiol B*, 115 (2012) 9–15.

14. Ghojavand H, Vahabzadeh F, Roayaei E & Shahraki AK, Production and properties of a biosurfactant obtained from a member of the Bacillus subtilis group (PTCC 1696), *J Colloid Interface Sci*, 324(2008) 172–176.

15. Gudina EJ, Teixeira JA & Rodrigues LR, Isolation and functional characterization of a biosurfactant produced by *Lactobacillus paracasei*, *Colloids Surf*, 76 (2010) 298–304.

16. Persson A, Osterberg E & Dostalek M, Biosurfactant production by Pseudomonas fluorescens 378: growth and product characteristics, *Appl Microbiol Biotechnol*, 29(1988) 1–4.

17. Abouseoud M, Maachi R, Amrane A, Boudergua S & Nabi A, Evaluation of different carbon and nitrogen sources in production of biosurfactant by *Pseudomonas fluorescens*, *Desalination*, 223 (2008) 143–151.

18. Mulligan CN & Gibbs BF, Correlation of nitrogen metabolism with biosurfactant production by *Pseudomonas aeruginosa*, *Appl Environ Microbiol*, 55 (1989) 3016–3019.

19. Abu-Ruwaida AS, Banat IM, Haditirto S, Salem A & Kadri M, Isolation of biosurfactant producing bacteria, product characterization, and evaluation, *Acta Biotechnol*,11(1991) 315–324.

Table 1- Data of cellular growth and production of biosurfactant (acetone extraction)

Time (Hrs)	Biosurfactants (mgL^{-1})	Biomass (g L^{-1})
1	11.83	0.11
2	21.63	0.116
4	50.34	0.121
6	63.23	0.203
8	89.46	0.206
10	175.30	0.213
20	243.62	0.767
30	249.30	2.243
40	253.79	3.564
50	113.12	3.462
60	100.22	3.423
70	98.33	3.29
80	95.22	3.30

Table 2- Amino acid analysis

Sr.No.	Test	Observation	Results
1	Physical appearance of solution	Clear	Peptone
2	Color	Faint yellow	Peptone
3	Litmus test	Acidic	Peptone
4	Biurret test O.S.+2ml NaOH +2-3 drops of CuSO4	Violet ring	Protein present

Table 3- Functional group and peak frequency of IR spectra

Functional Group	Peak frequency (cm^{-1})
Peptides	3188
NH-stretching	1652
–CH2–of a carbonyl group	2927-2959
O-H stretching	3307
Alkane group	1388

Fig.1 (a-c) – Biosurfactant production [a. Fermentation medium before centrifugation; b. Cell free supernatant after centrifugation & c. Isolated biosurfactant from supernatant]

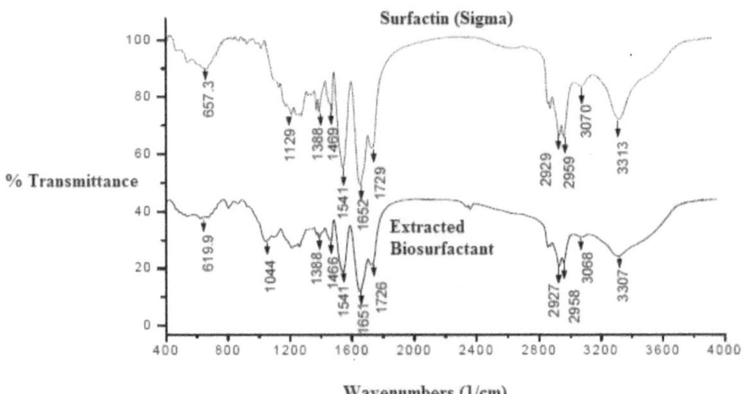

Fig. 2- IR spectra of Surfactin (Sigma Aldrich) and isolated Biosurfactant

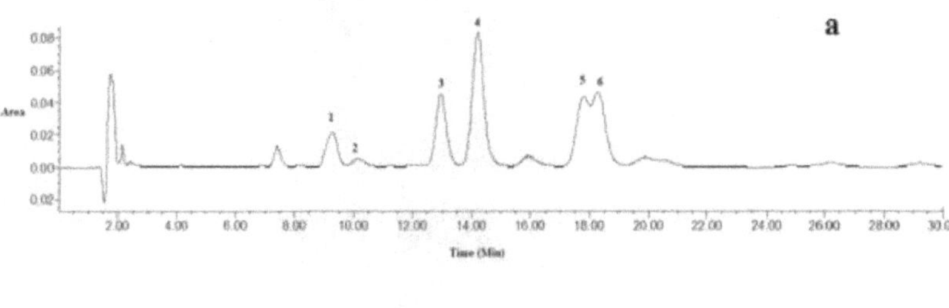

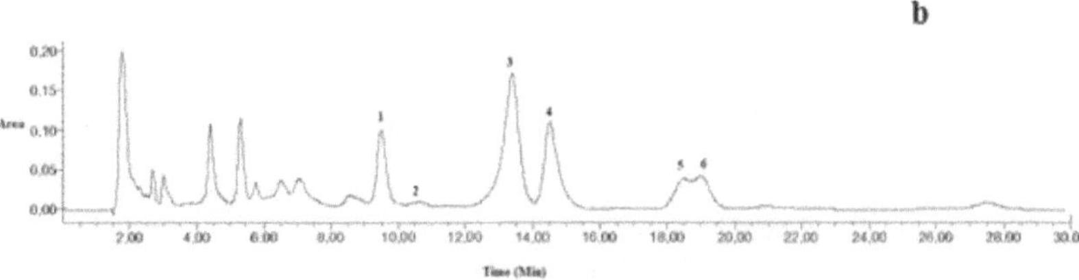

Fig.-3 HPLC chromatogram of [a. marketed surfactin (Sigma); b. extracted biosurfactant.

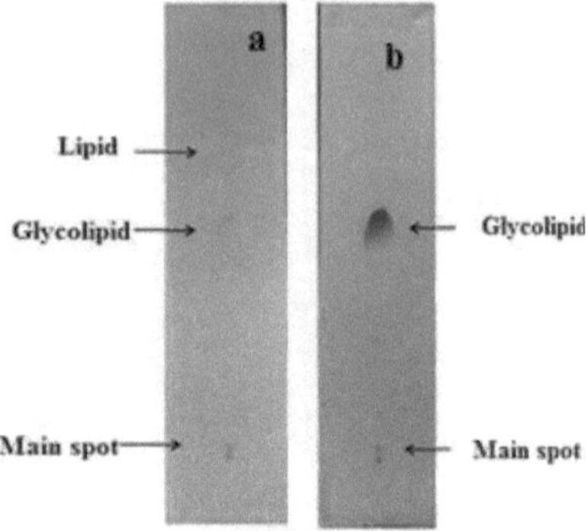

Fig. – 4 TLC study of Biosurfactant [a. isolated biosurfactant; b. Surfactin (Sigma Aldrich)]

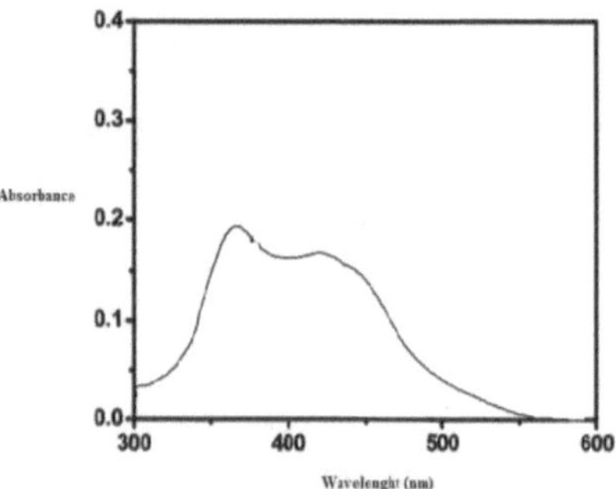

Fig.-5 Absorbance spectra of curcumin (30ppm) at CMC in extracted biosurfactant solution

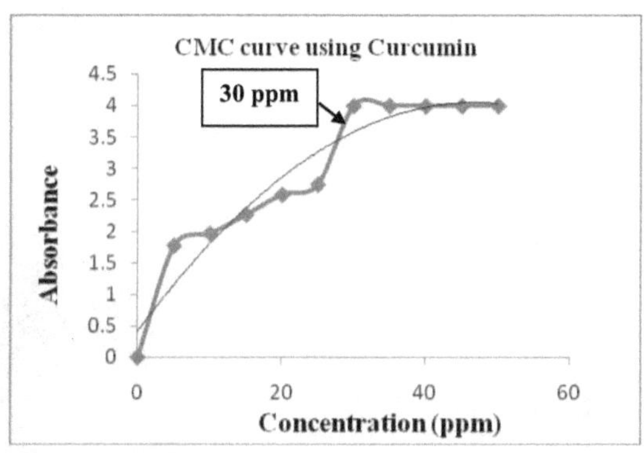

Fig. 6- CMC curve of isolated surfactant.

YOUR KNOWLEDGE HAS VALUE

- We will publish your bachelor's and master's thesis, essays and papers

- Your own eBook and book - sold worldwide in all relevant shops

- Earn money with each sale

Upload your text at www.GRIN.com and publish for free